ATOMS

Rebecca Woodbury, Ph.D., M.Ed.

Gravitas Publications Inc.

ATOMS

Illustrations: Janet Moneymaker
Design/Editing: Marjie Bassler, Rebecca Woodbury, Ph.D., M.Ed.

Atoms
ISBN 978-1-950415-09-0

Published by Gravitas Publications Inc.
Imprint: Real Science-4-Kids
www.gravitaspublications.com
www.realscience4kids.com

RS4K

Image credits: p.3. Дмитрий Ларичев, iStock; p. 5. Roaj Kanishthanaka, iStock; p. 7. Heru Agung, iStock

Is the moon made of green cheese?

Hmm...
I don't think so.

Are clouds made
of cotton candy?

That would
be yummy!

Are frogs made of green slime?

Yes!
I think so.

The moon is not made of green cheese.
The moon is made of **atoms**.

Clouds are not made of cotton candy.
Clouds are made of **atoms**.

Frogs are not made of green slime.
Frogs are made of **atoms**.

Oh! They are all made of atoms.

Atoms are like little building blocks.

Atoms can link together to make moons, clouds, and frogs.

Atoms make everything we touch, taste, smell, or see.

Wow! Atoms are cool!

Sodium

Oxygen

Hydrogen

We are building blocks.

Atoms have different names.

Those are funny names!

In this book we show atoms as drawings with arms and hands.

An atom does not really have arms and hands.

Drawing arms and hands helps us understand how atoms work.

I have arms and hands!

Phosphorus

Carbon

Nitrogen

Oxygen

Each arm drawn on an atom stands for one electron.

Electrons can link atoms together.

These electrons can be called **linking electrons**.

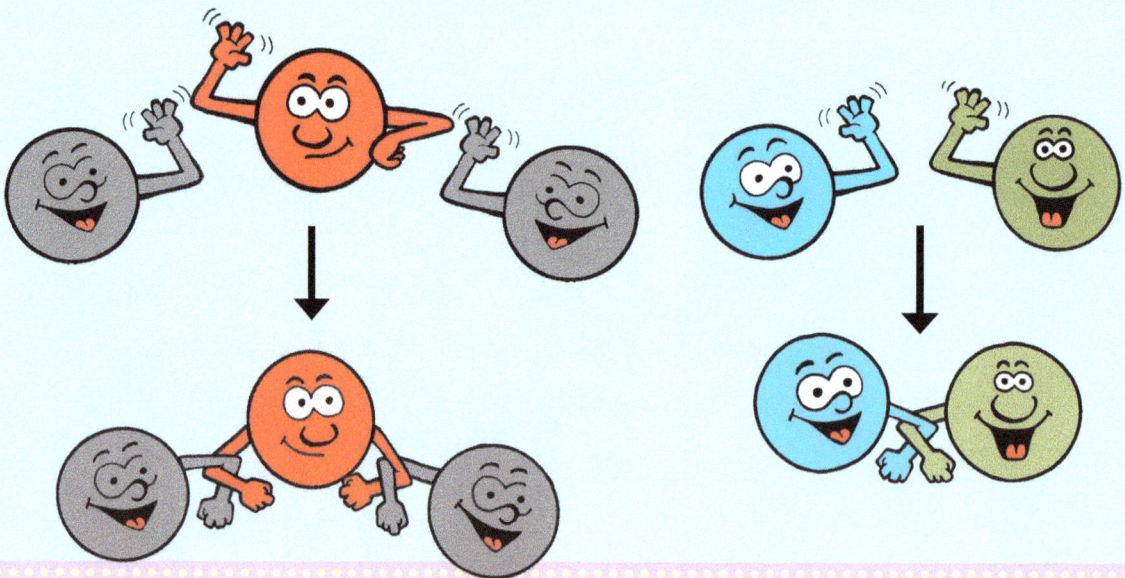

Different atoms have different numbers of linking electrons.

We show this by drawing atoms with different numbers of arms.

I have **four linking electrons**.

I have **one linking electron**.

Carbon

Chlorine

How many **linking electrons** does **phosphorus** have?

Hint. Count the number of arms.

My name is **phosphorus**.

How many **linking electrons** does boron have?

Remember the hint!

How to say science words

atom   (AA-tum)

boron   (BAWR-ahn)

carbon   (KAR-buhn)

electron   (i-LEK-trahn)

hydrogen   (HIY-druh-juhn)

nitrogen   (NIY-truh-juhn)

oxygen   (AHK-sih-juhn)

phosphorus   (FAHS-fuh-rus)

sodium   (SOH-dee-uhm)